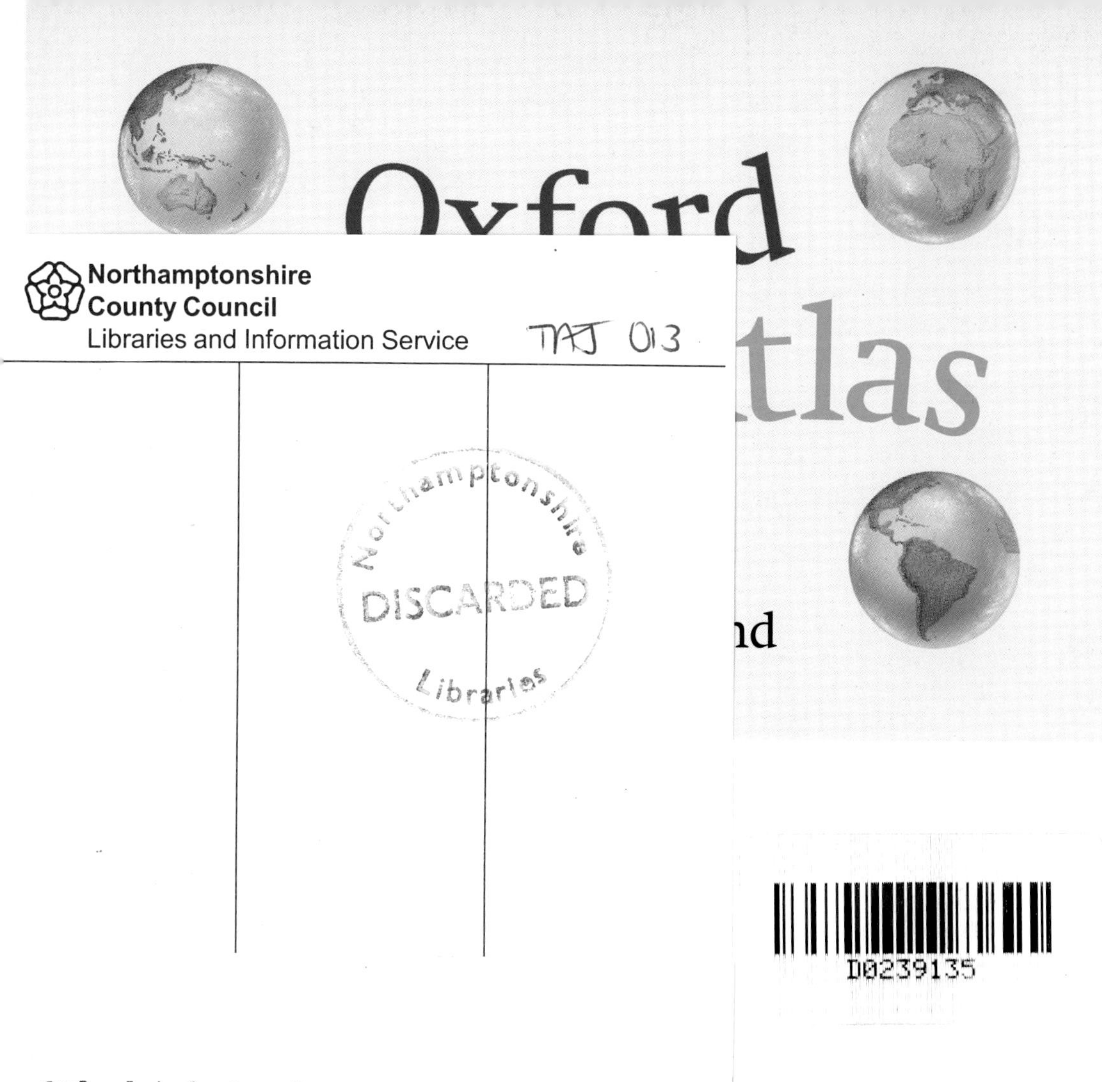
Oxford
atlas
nd



**Acknowledgements**

Illustrations: Bob Chapman (views of the Earth); Tessa Eccles (globes, maps); Oxford Illustrators (flags); Carleton Watts (space backgrounds).

Photog ... Arnold);
...

80 002 558 706

# Contents

## The Earth

## Maps of the world



## Europe

## The British Isles



## The United Kingdom (UK)



## Index 32

**A**
Africa 17
Alps 15, 22
Andes 14
Antarctica 16,17
Arctic Ocean 14, 15
Asia 17
Atlantic Ocean 14, 20, 22, 27
Australia 19

This is the Earth in space.

 The Earth is round, like a ball.

 There is land and sea.

# A globe

A globe is a model of the Earth.

# Views of the Earth

 Each view of the Earth is different.

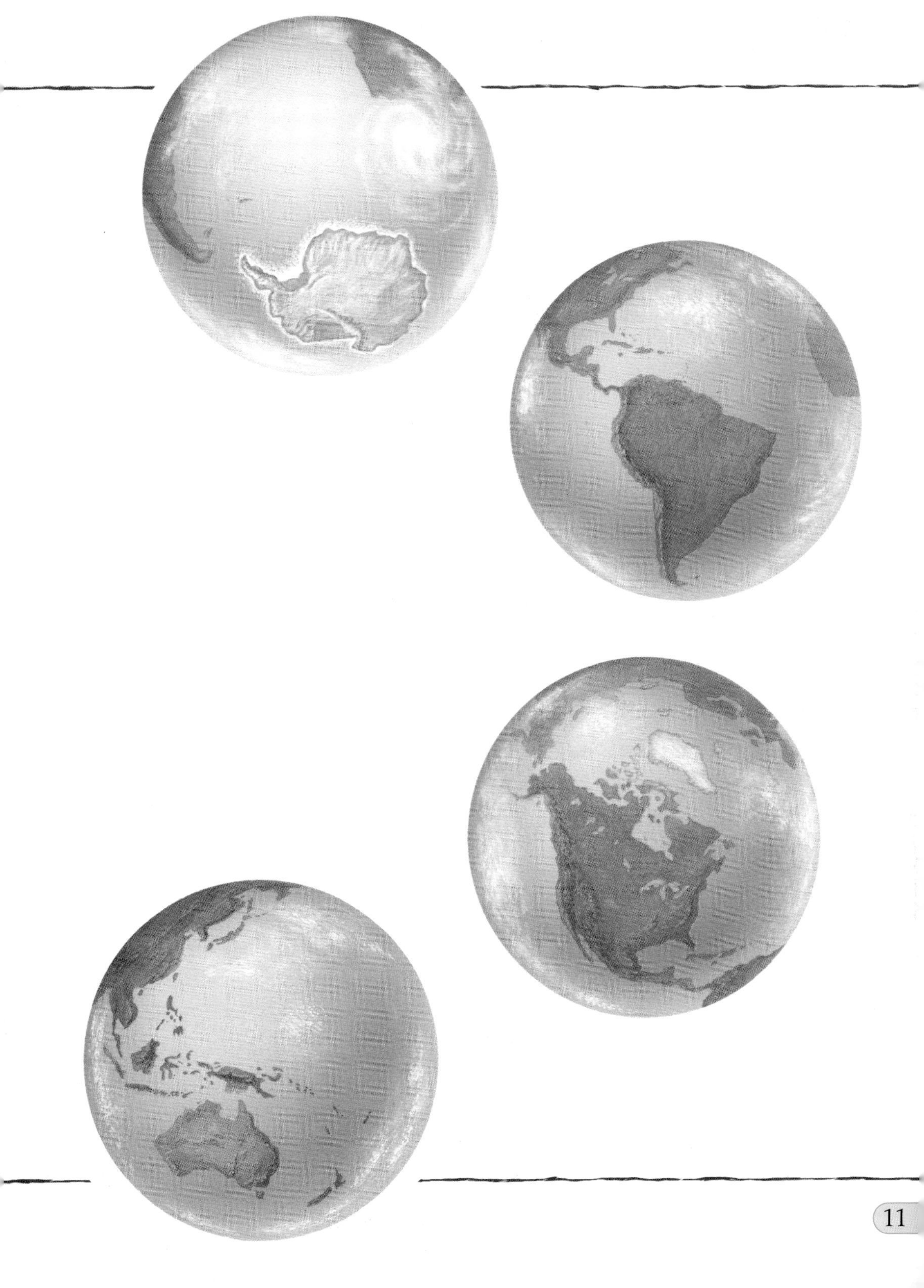

# The world

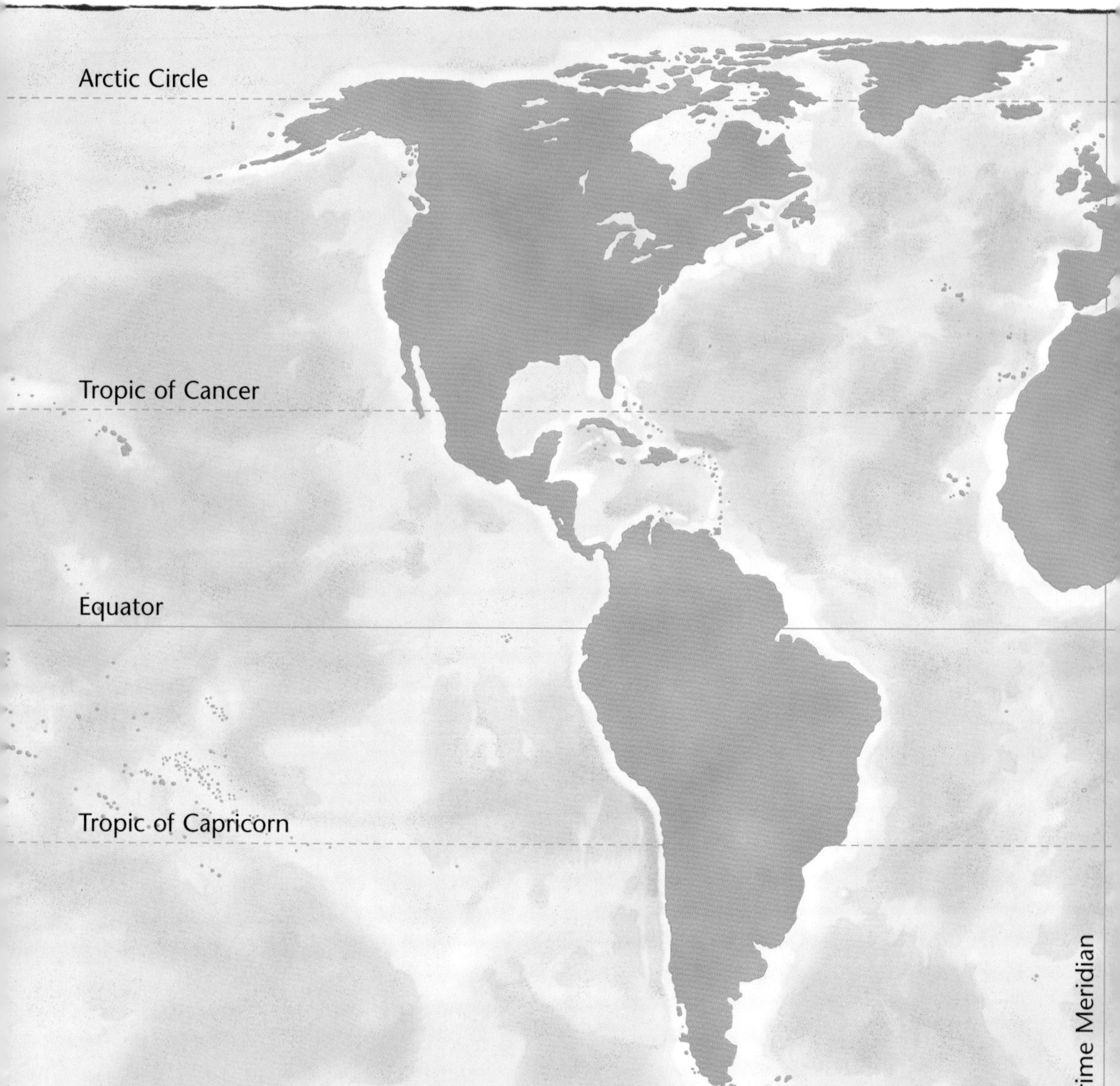

Arctic Circle

Tropic of Cancer

Equator

Tropic of Capricorn

Antarctic Circle

Prime Meridian

 This is a map of the world.

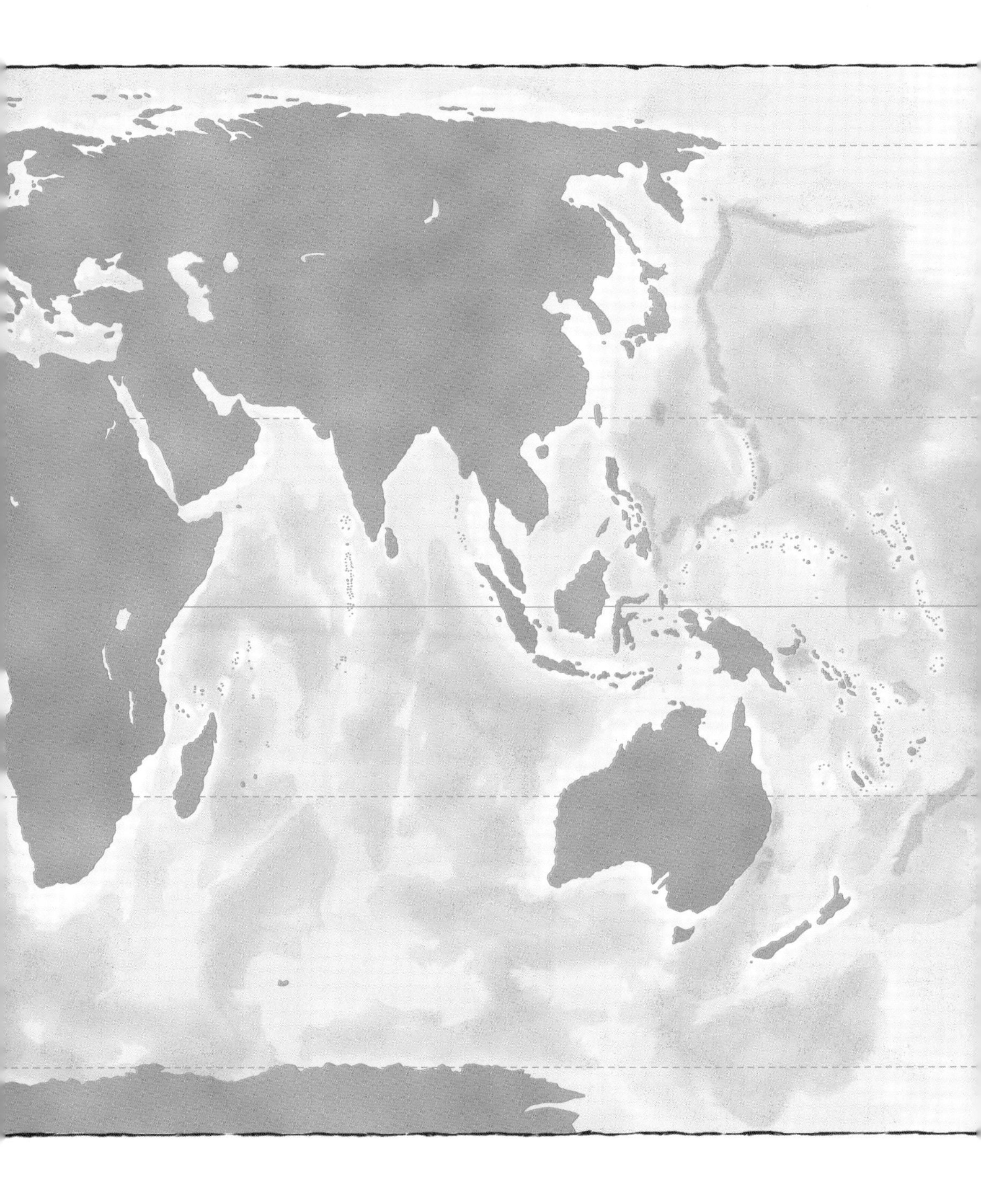

**Key** land sea

# The world

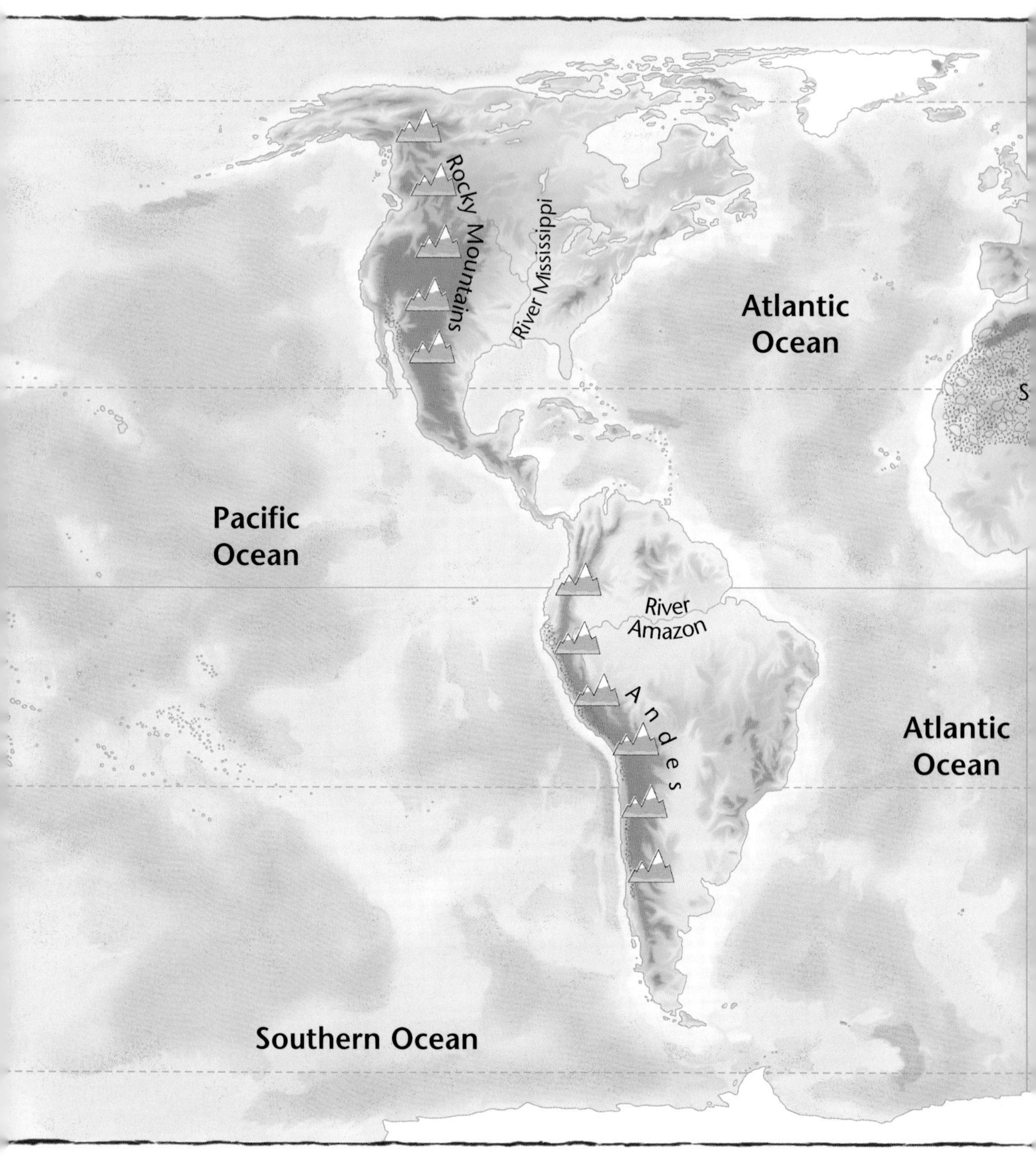
Rocky Mountains
River Missisippi
Atlantic Ocean
Pacific Ocean
River Amazon
Andes
Atlantic Ocean
Southern Ocean

 There are rivers, mountains, and deserts.

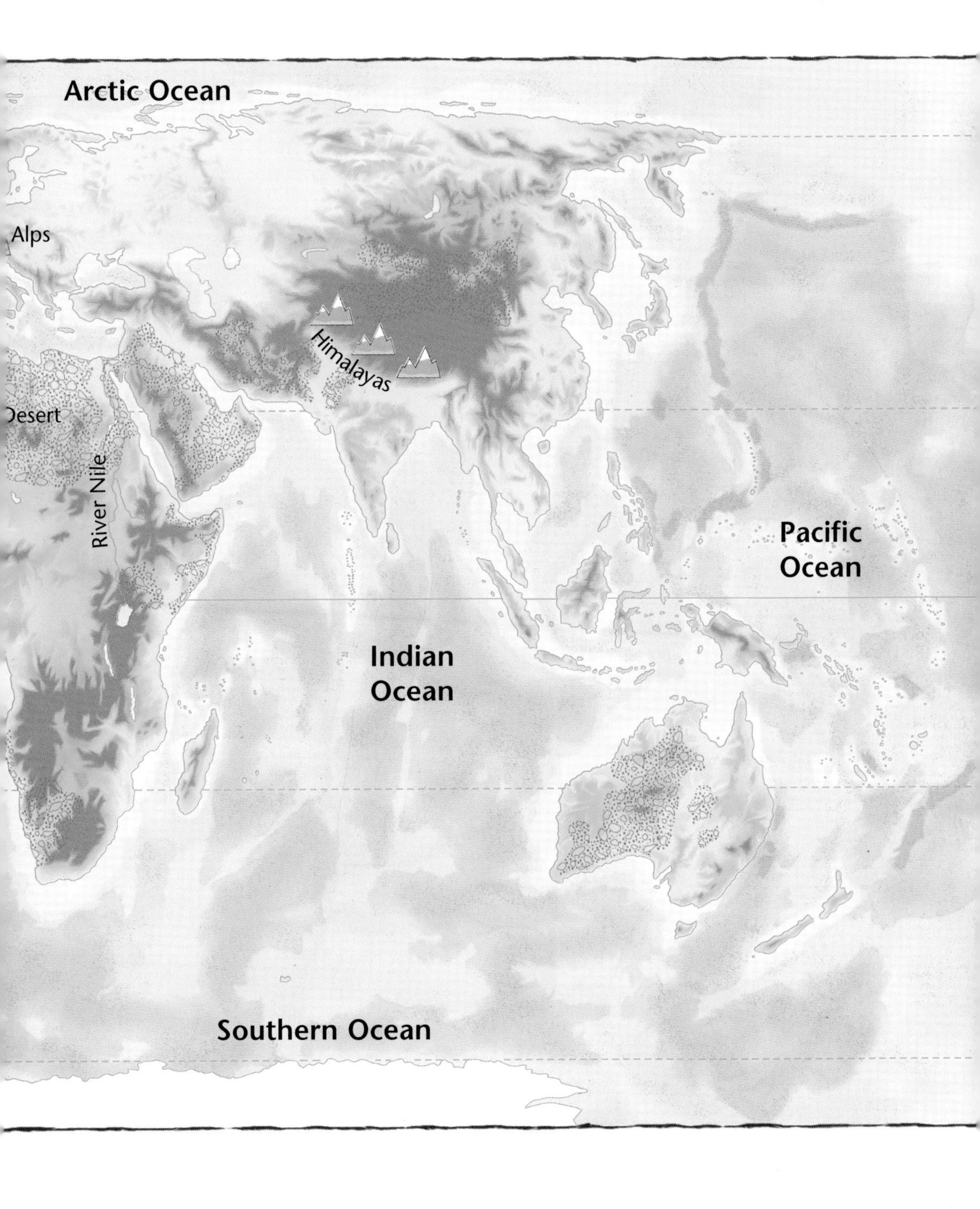
Arctic Ocean
Alps
Himalayas
Desert
River Nile
Pacific Ocean
Indian Ocean
Southern Ocean

Key  rivers mountains 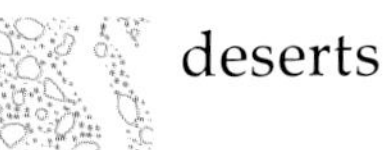 deserts

# The world

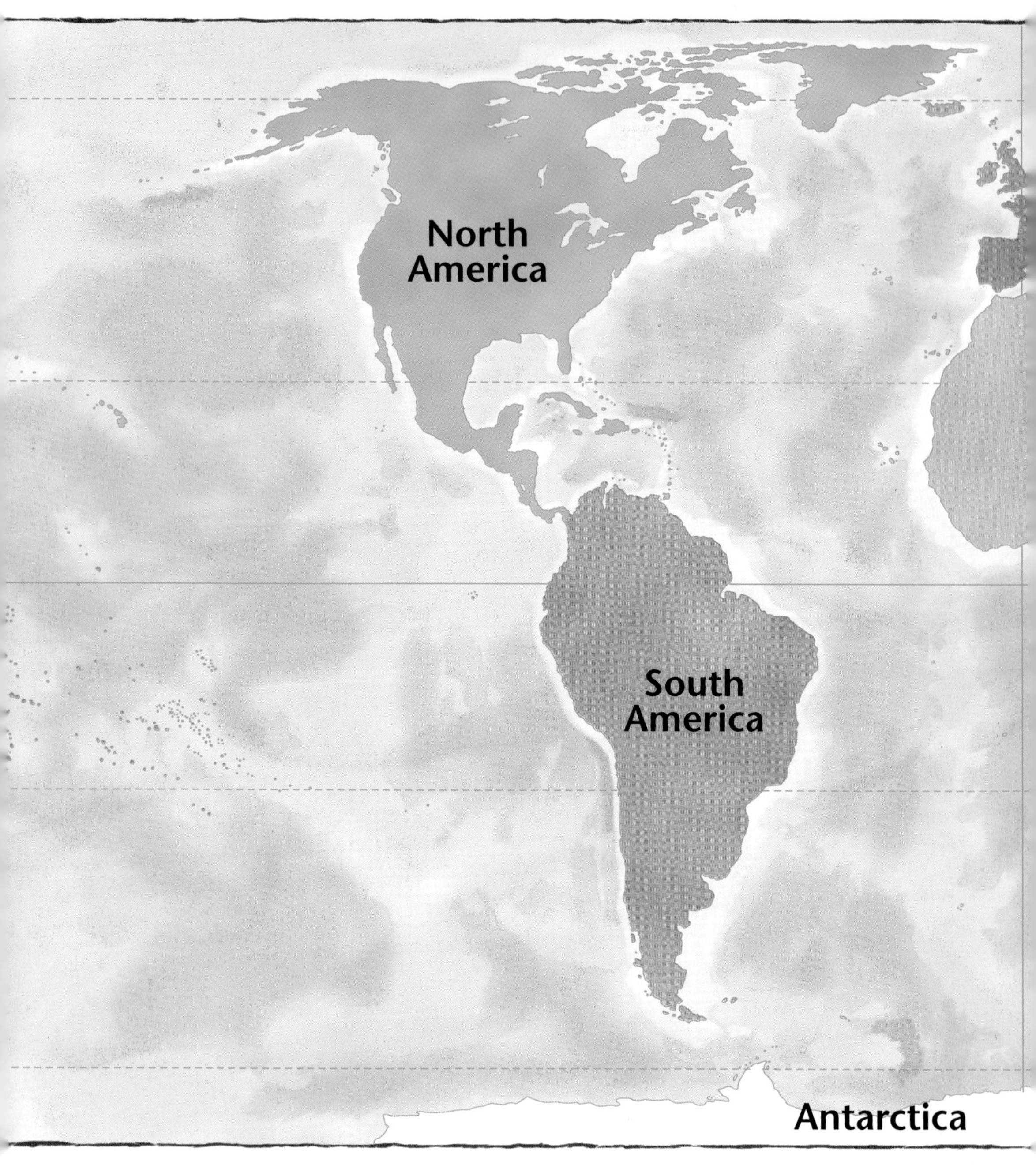
North America
South America
Antarctica

There are seven continents.

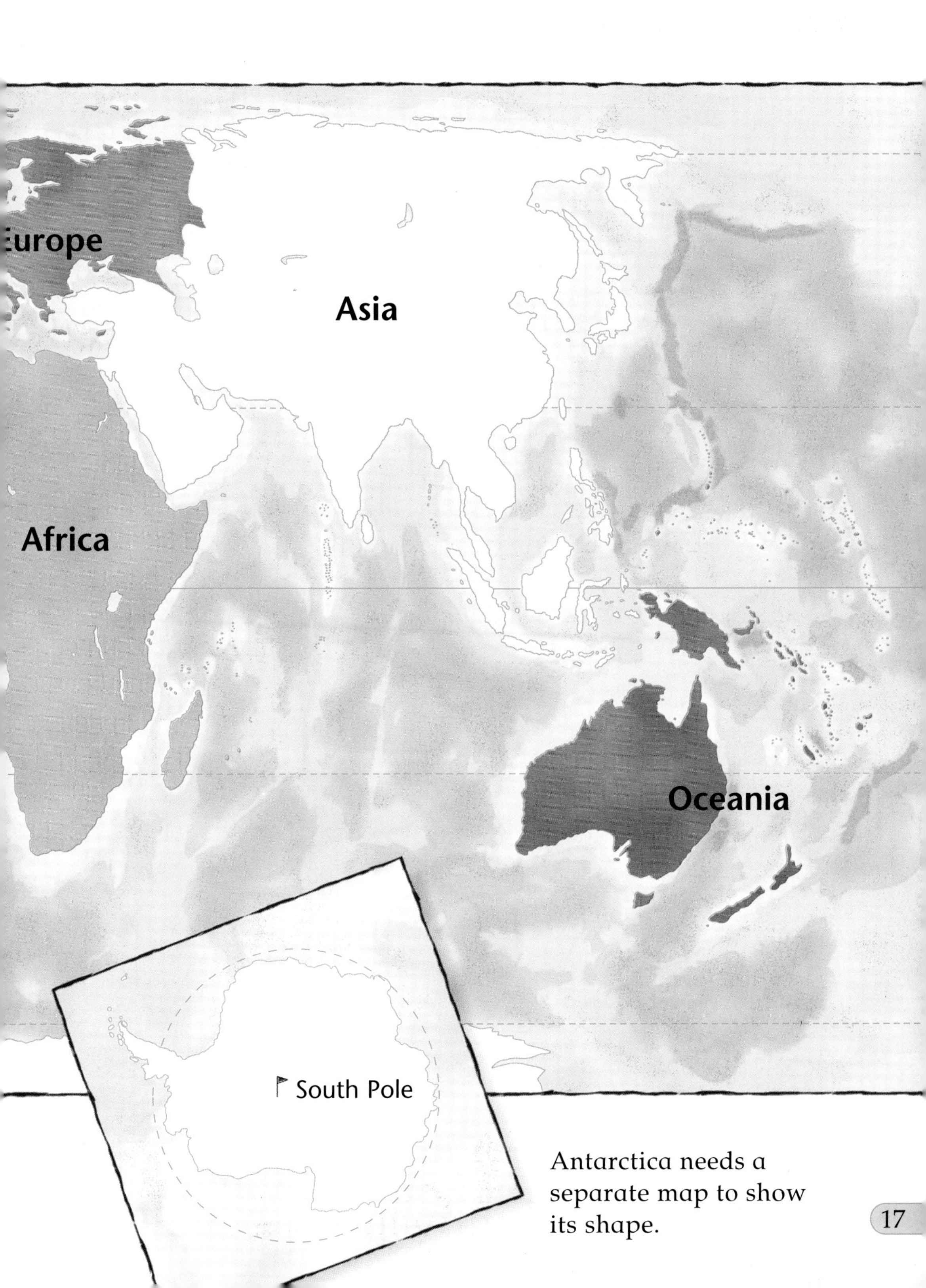
Europe
Asia
Africa
Oceania
South Pole

Antarctica needs a separate map to show its shape.

# The world

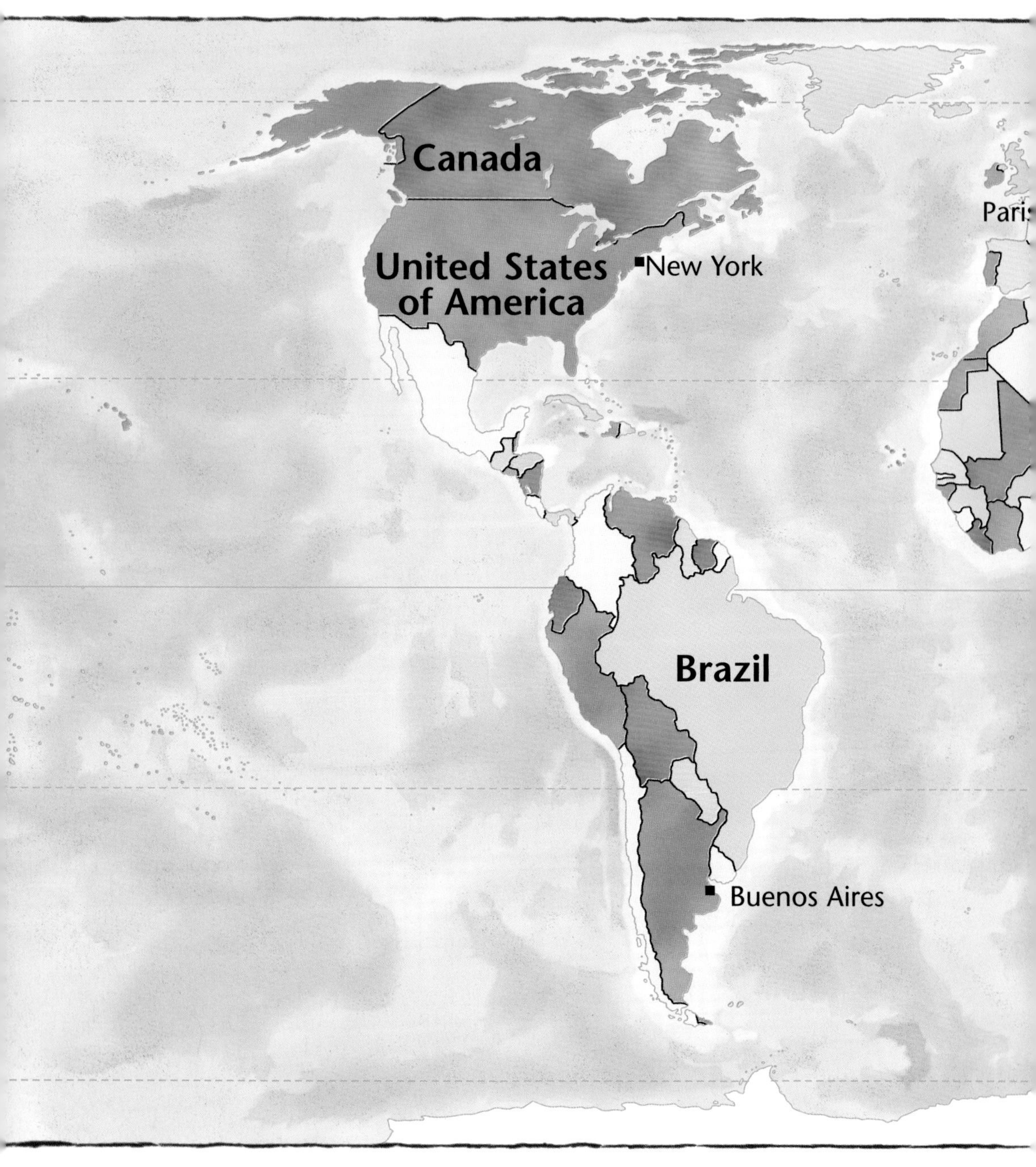
Canada
United States of America
New York
Brazil
Buenos Aires

 There are many countries and cities.

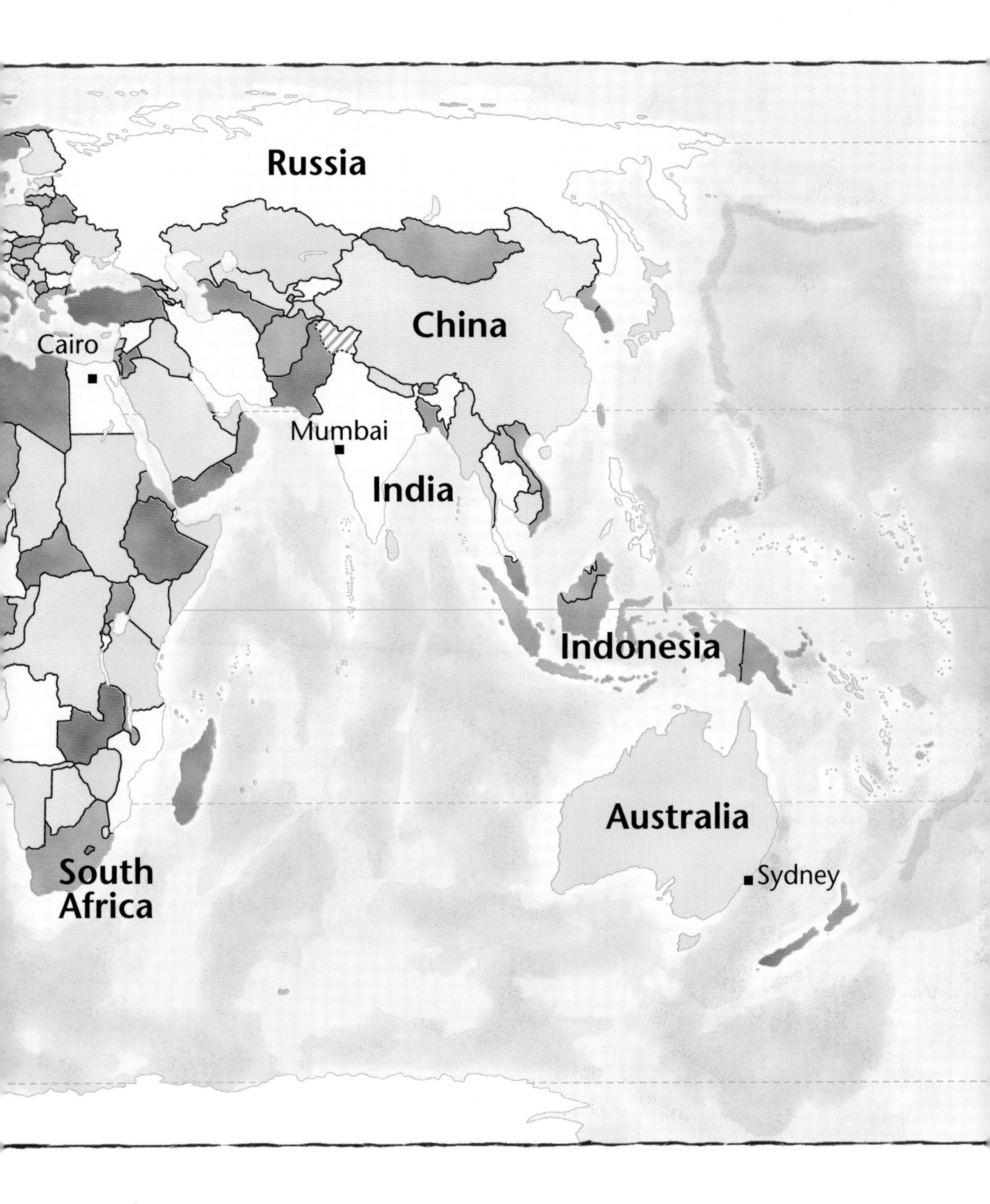
Russia
China
Cairo
Mumbai
India
Indonesia
Australia
Sydney
South Africa

**Key** Colours show countries, Some are named on the map. ■ some big cities

# Europe

Arctic Circle

Prime Meridian

British Isles

Atlantic Ocean

Eu

Mediterranea

Europe is the smallest continent.

ope
ea
Key
land
sea

# Europe

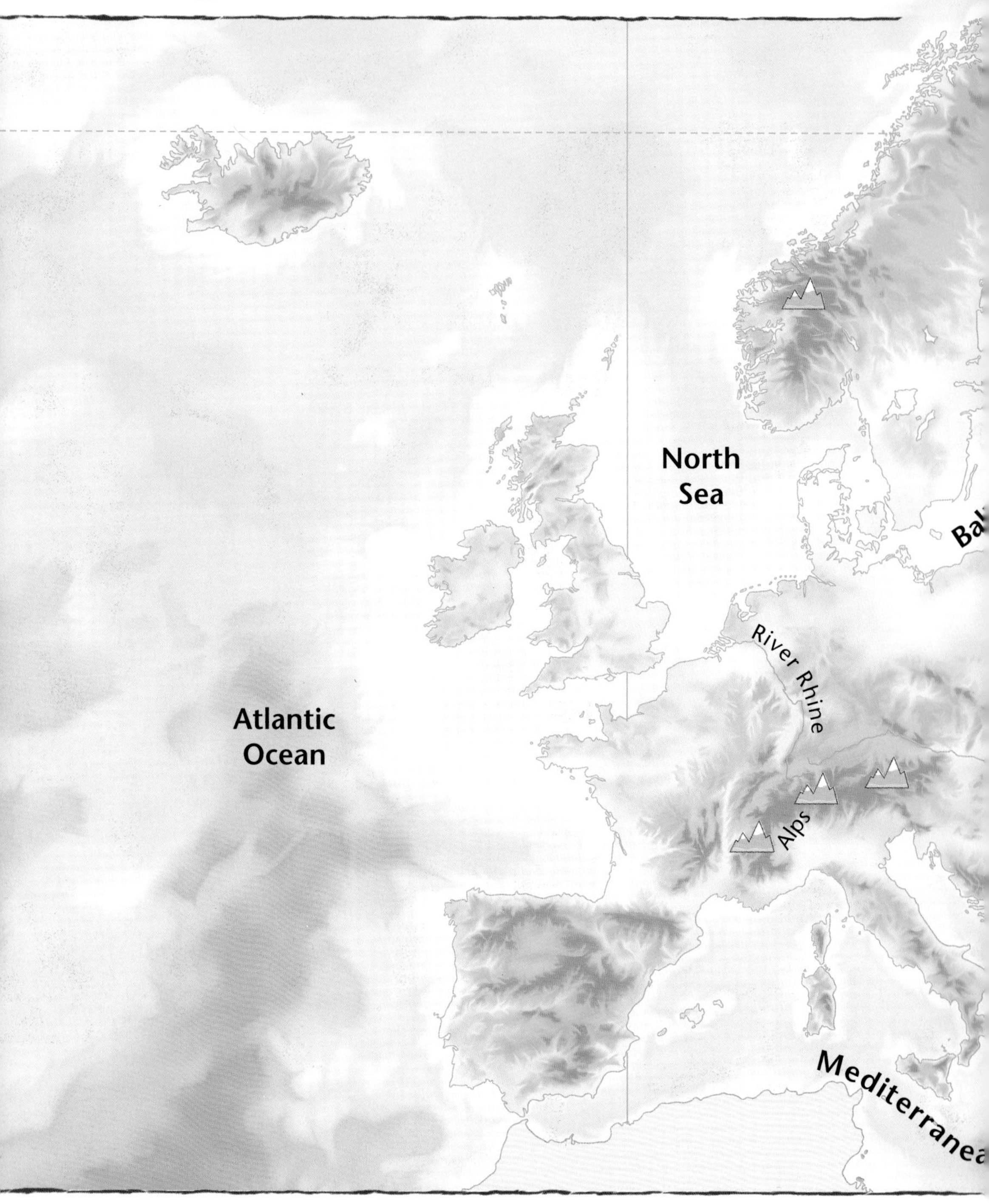
North Sea
Atlantic Ocean
River Rhine
Alps
Mediterranea

 There are rivers and mountains.

River Danube
Sea
Key
rivers
mountains

# Europe

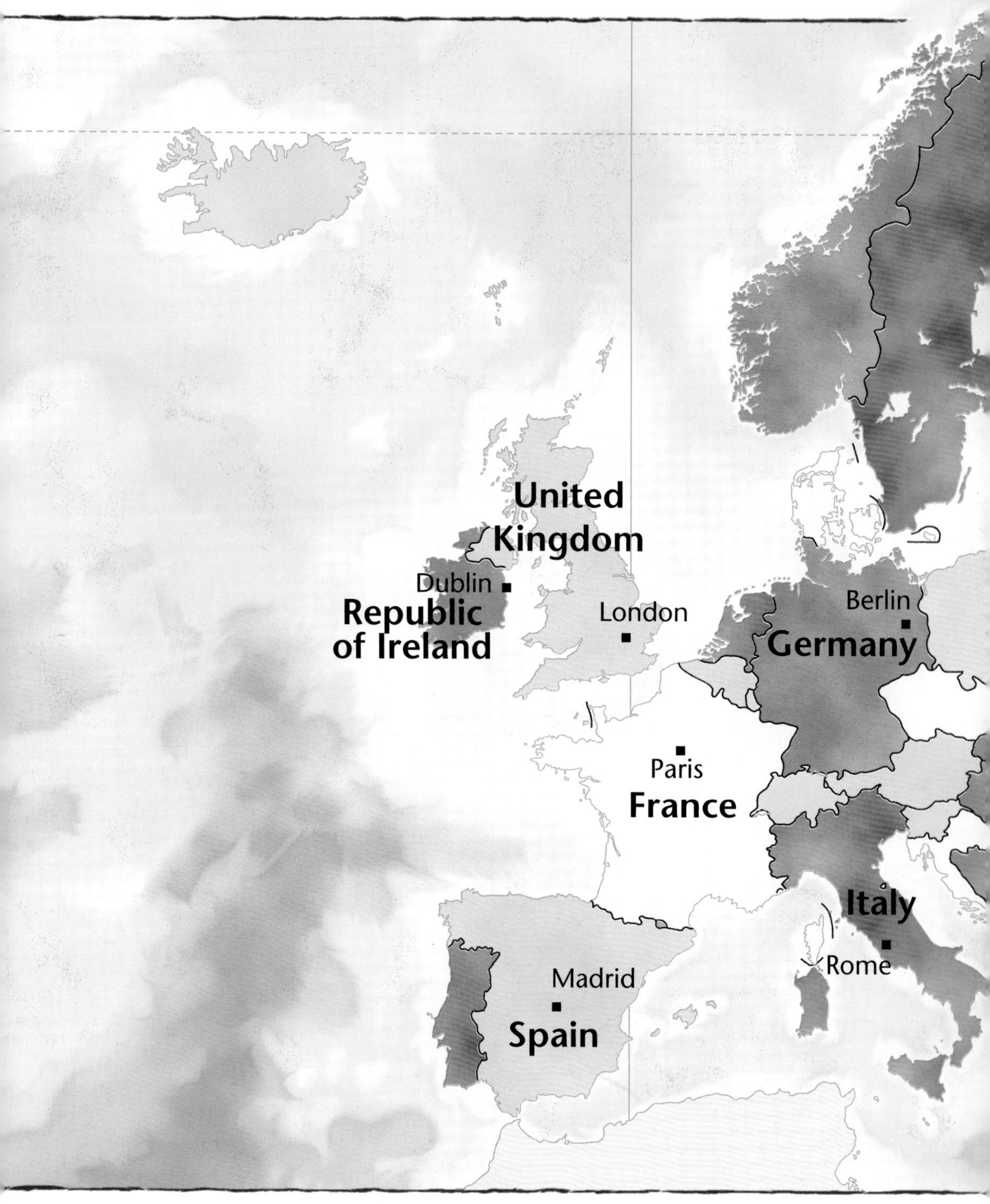
United Kingdom
Dublin
Republic of Ireland
London
Berlin
Germany
Paris
France
Italy
Rome
Madrid
Spain

 There are many countries and cities.

Russia

**Key** Colours show countries. Some are named on the map.

■ capital cities

# The British Isles

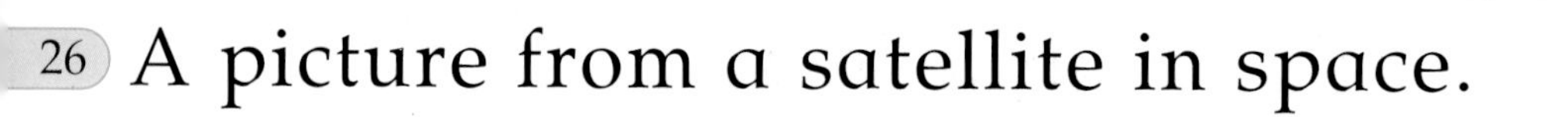

 A picture from a satellite in space.

# The British Isles

Key
land
sea

Prime Meridian

North Sea

Ireland

Irish Sea

Great Britain

Atlantic Ocean

English Channel

There are two large islands.

# The British Isles

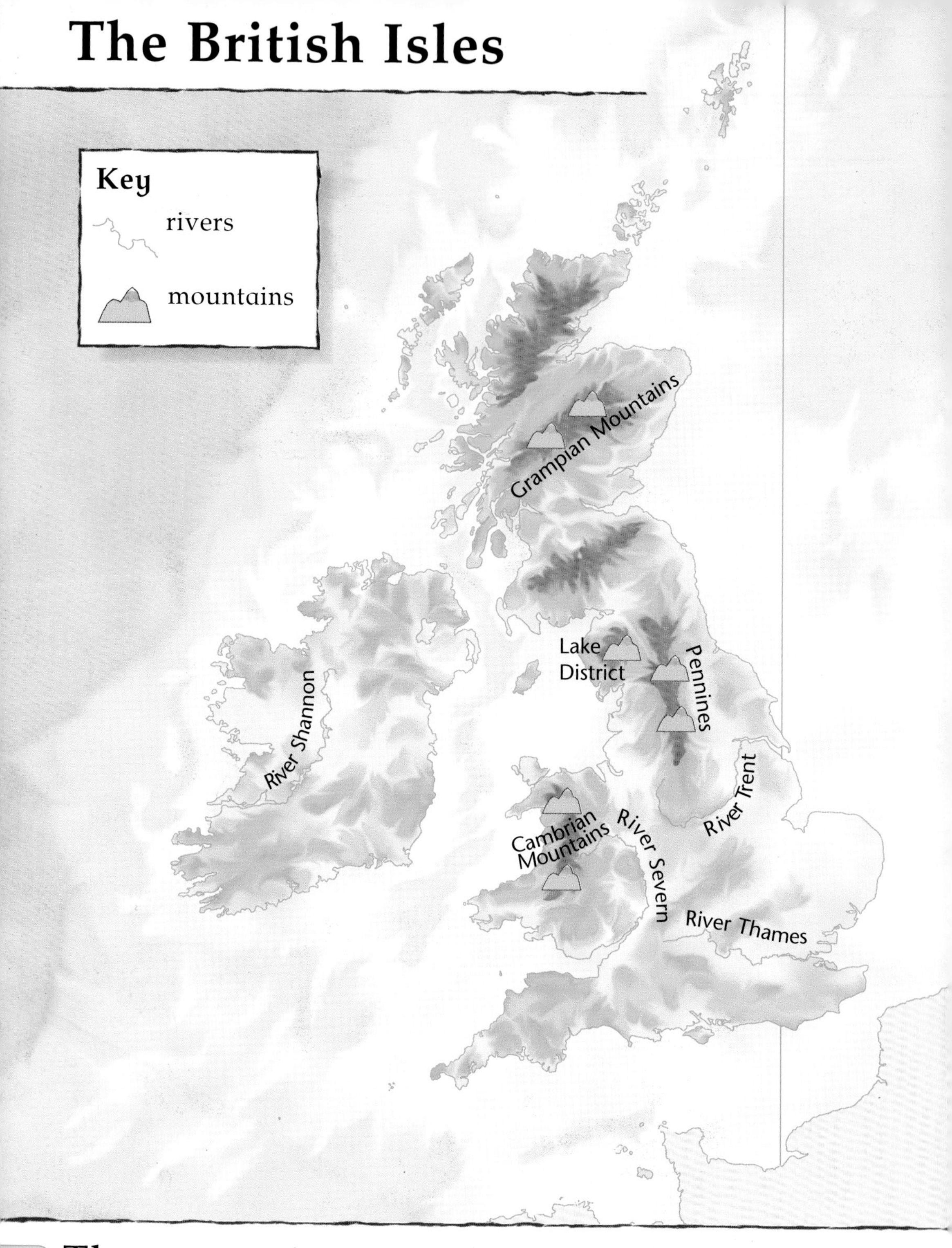
Key
rivers
mountains
Grampian Mountains
Lake District
Pennines
River Shannon
River Trent
Cambrian Mountains
River Severn
River Thames

There are rivers and mountains.

# The British Isles

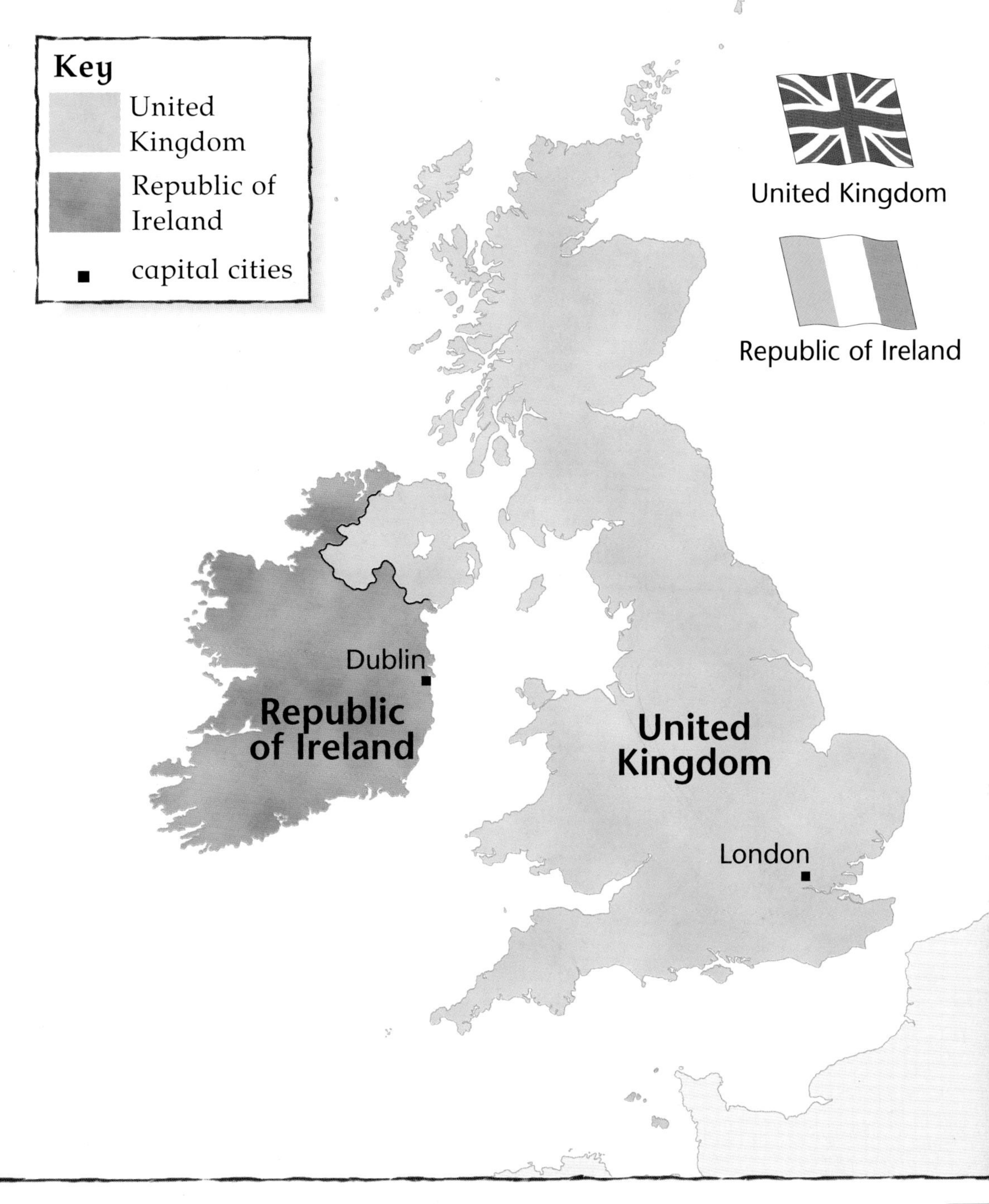
Key
United Kingdom
Republic of Ireland
capital cities
United Kingdom
Republic of Ireland
Dublin
Republic of Ireland
United Kingdom
London

There are two countries.

# The United Kingdom

Key
England
Scotland
Wales
Northern Ireland
Scotland
Northern Ireland
Wales
England
Scotland
England
Wales

 The United Kingdom has four parts.

# The United Kingdom

**Key**

- ■ capital cities
- • other big cities

Glasgow
Edinburgh
Newcastle upon Tyne
Belfast
Leeds
Liverpool
Manchester
Sheffield
Nottingham
Birmingham
Norwich
Cardiff
London
Bristol

There are many big cities.

# Index

name of place

London 24, 29, 31

the pages where you will find it





A list of place names in this atlas.